The Discovery Area

By Evelyn Ayum

Library of Congress Control Number: 2016921352 Essentials by Evelyn Publishing Bayonne, New Jersey. All rights are reserved. Written permission from the author can be obtained in writing. Ebolocs@aol.com

The Discovery Area is about stimulating children interest to explore in the discovery area of the classroom. This area should spark children's curiosity into the world around them. Children should be excited to explore the many objects such as, leaves, coins, shells, beans, telescopes, magnifying glasses, binoculars, bird seeds, pets, bones, rocks and a vast amount of things in science. The discovery area can lead children to be inquisitive about nature, the structure of the human body and animals from the past and present. It is the teacher's job to encourage and provide children with the necessary tools to imagine and discover the world outside. I hope you enjoy this book and have fun in the discovery area.

Evelyn Ayum

Library of Congress Control Number: 2016921352 Essentials by Evelyn Publishing Bayonne, New Jersey. All rights are reserved. Written permission from the author can be obtained in writing. Ebolocs@aol.com

The discovery area is a place to explore to look at things up close, to see the world from different perspective up and down and all around.

Pine cones, leaves, seeds and weeds and most anything that grows from the ground and feed the people to get fat and round.

Butterflies and insects, bumble bees, buzzing all around dropping nectar here and there and everywhere. Learn how this happens in the discovery area

Magnifying glasses, magnets shaped like horse shoes, rings that attract to all metals and key rings. This is what I discovered in the discovery area.

Microscopes to see small objects, a single hair strand on my head or a tiny bug on my bed. I can use this tool to magnify anything that I can't see with my naked eyes.

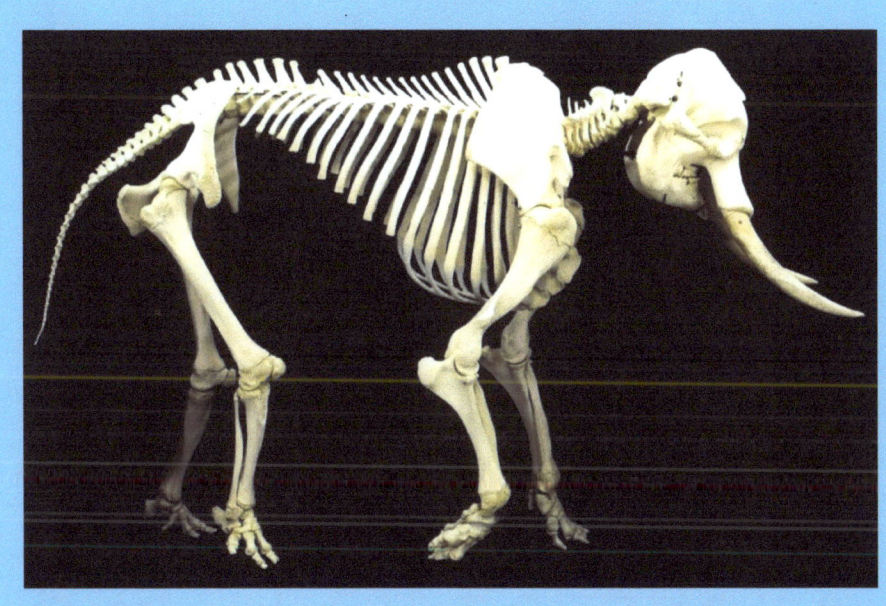

The skeleton of our body using an x-ray to see our bones, legs, arms and head and even our cheek bones can be seen with a special machine.

Discover the view of the earth from up above the sky and clouds, far away up, up away high until you reach out of space.

Discover the stars and planets, the moon so bright and the orbit of the sun as it warms the earth. You can discover it here in the discovery area where your imagination can soar.

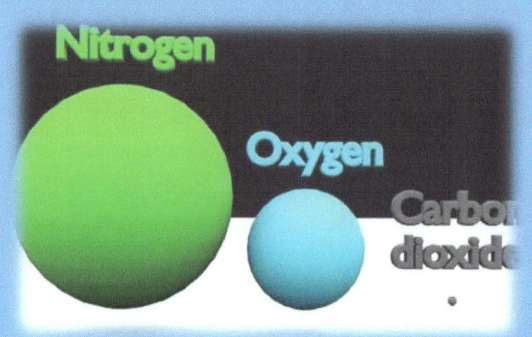

Oxygen, nitrogen and carbons all make up the elements of the air we breathe and the tree make it for us oxygen to breathe.

We watch and observe the cycle of life and discover how animals and insects do the same.

A frog grows from a tadpole to a frog and this cycle repeats itself again and again.

A butterfly grows from an egg
to a caterpillar then Chrysalis to
and an adult and repeat this
cycle again and again.

A human grows from an embryo and gives birth and develops from a toddler to a full grown adult and the cycle continues. All this I discovered in the discovery area.

Discover the many scientific
things to do like mixing colors
of varying hues to cutting up
things like triangles and cubes.

Listen to the sea shell and the many sounds that objects make, or sort out the ingredients to make a cake. All of this I learned in the discovery area.

Create a volcano out of clay using baking soda, flour, salt, vinegar, food coloring and more. Watch it bubble over all over the table, chair and floor.

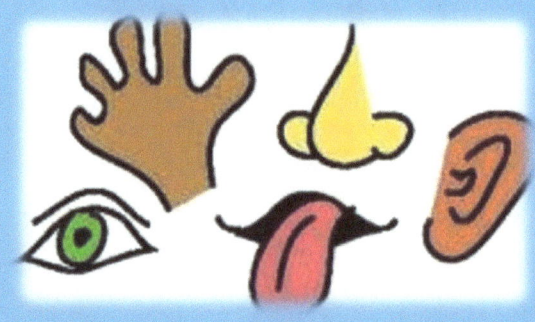

You can use your five senses for so many things. Your nose to smell the different scents, your ears to hear, fingers to touch and feel the many texture of the world and the tongue tastes the foods that we all love. You can discover all of this in the discovery area.

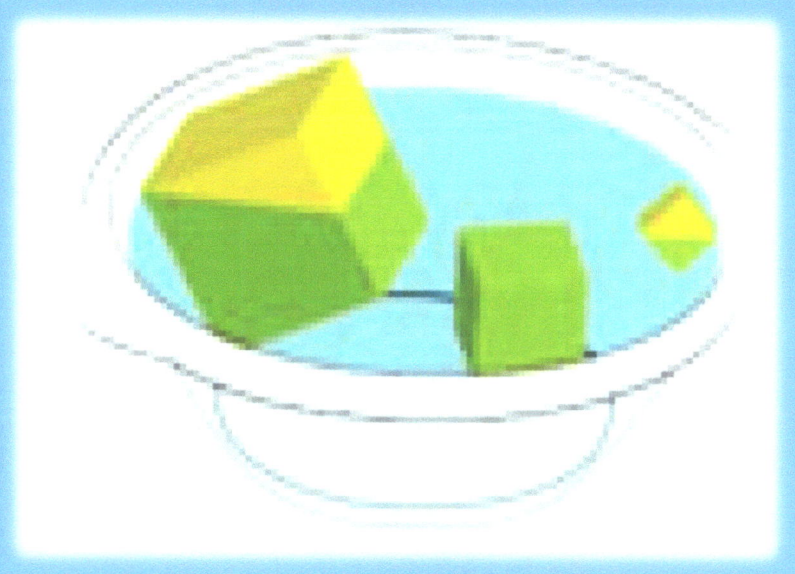

Make predictions as you float
and sink objects of all kinds and
discover how water changes
things when wet and then
watch them dry.

Discover the many ways to create a rose, airplane and snowflake from paper that you fold.

warm up oven to 400° F

mix together:

1¼ c flour
¾ c Corn Meal
¼ c sugar
2 teaspoons baking powder
¼ teaspoon salt

then stir in:

1 c milk
¼ c olive oil
1 egg (beaten)

pour batter into a greased pan, and bake for 25 minutes. The pan can be a 8" or 9" baking pan. My favorite is a pan with corn cob shapes. If I use it, I don't bake them as long.

Discover the ingredients and use recipes on various box tops and mix them up from batter to a cake. All this I discovered in the discovery area.

Discover the minerals from gold
to diamonds and all kinds of
rocks that the earth provides.

Discover the ways to recycle and reuse, to save and not waste whatever we use.

In the discovery area learn about space and the human race and the world we live in is a great place.

Discover the books in the discovery area that teaches us how to feed a nation as we learn to be citizens of a great nation. Growing things take time as we care for the plants and take care of the soil providing the earth with lots of

love.

The discovery area is a place to pretend using the binoculars to bird watch just me and a friend.

The discovery area has so many rocks of different kinds, colors, textures and shapes they are very fine to explore.

Discover the leaves and barks of trees or look at the labeled body structure of a flea.

When you are in the discovery area dream of what you can do and discover all the materials that you can use and reuse.

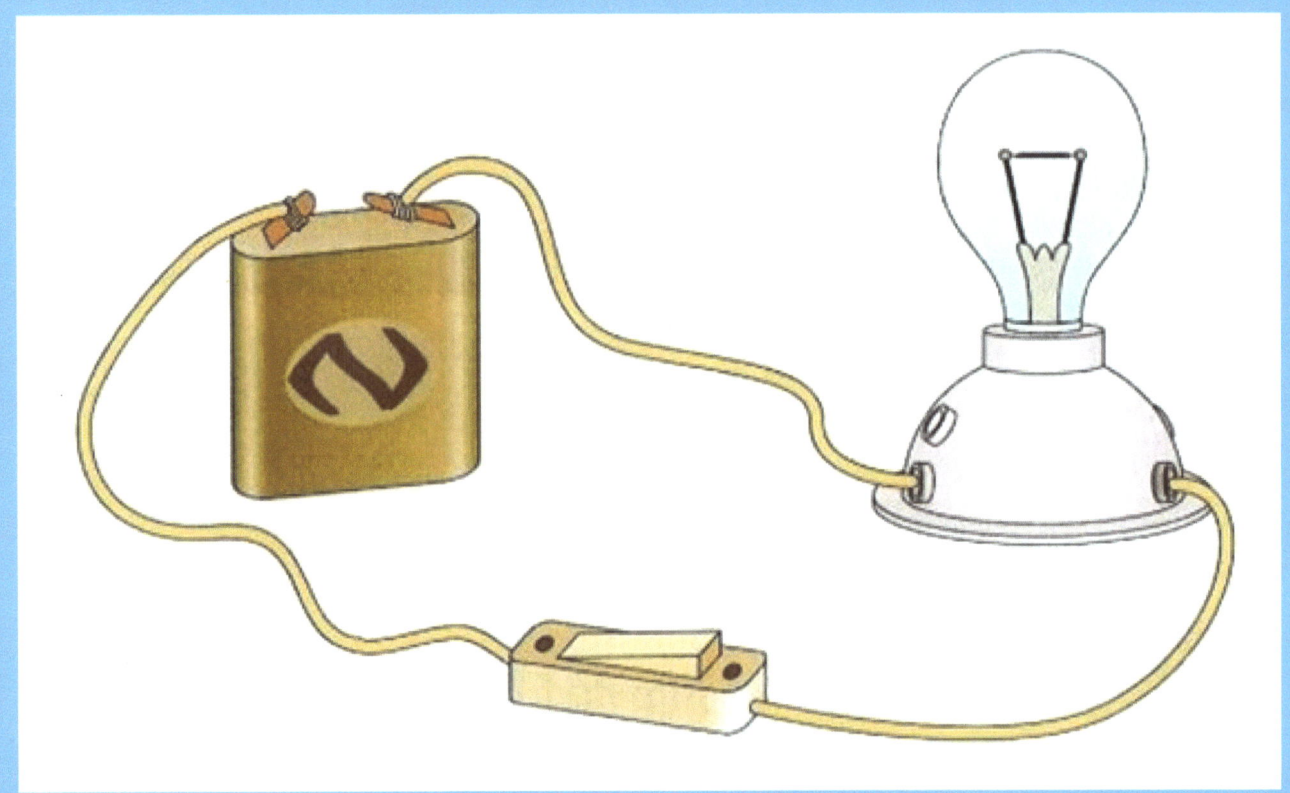

Discover how circuits work to light your home and try to create a switch of your own.

Learn to care for your pet and discover what a vet can do.

In the discovery area learn
about the earth and you. Learn
the many features that make
up people and you.

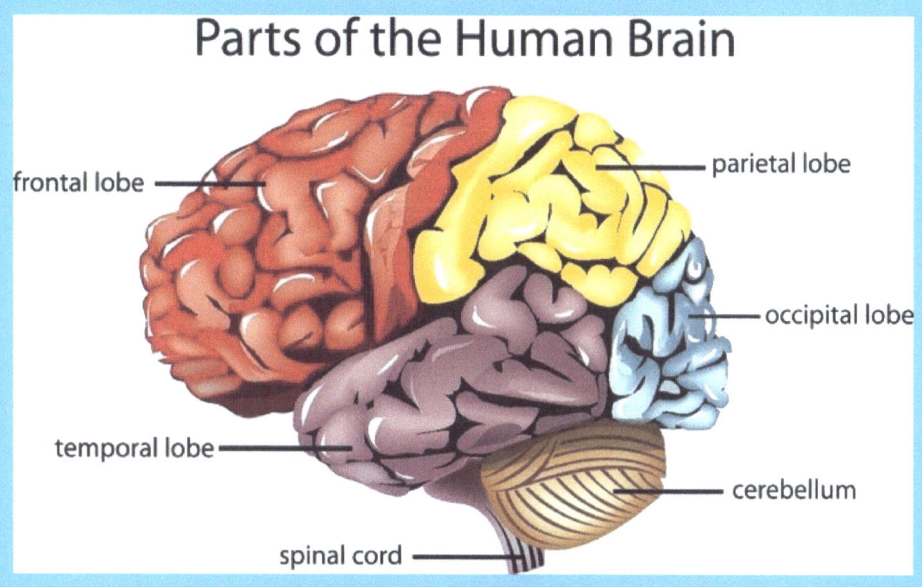

Discover parts of the brain and how human think and function using their brain.

There are all types of weather to discover, how to keep safe and secure in the storms this is the area you will learn all about this for sure.

Discover all the places to see using a map or a globe. You can pretend to visit places you've never been before.

Finally, in the discovery area use your imagination to figure out things to do. Work on projects together with a friend and you. Whatever you imagine you want to explore the discovery area will be waiting for you with lots more.

The beginning is not the end as you discover new explorations in the discovery area.

www.ingramcontent.com/pod-product-compliance
Lightning Source LLC
Chambersburg PA
CBHW050401180526
45159CB00005B/2107